Eliza
and the
Dragonfly

by Susie Caldwell Rinehart

illustrated by

Anisa Claire Hovemann

Dawn Publications

To my grandmothers, Marion and Teddy,
who were never afraid of something gross. — SCR

To the Creator for all the miraculous wonders
of this world. Also to my mom and dad,
to Garrett ❤ and to Hanna, Logan,
and the little one soon to come. — ACH

Copyright © 2004 Susan Caldwell Rinehart
Illustrations copyright © 2004 Anisa Claire Hovemann

A Sharing Nature With Children Book

Library of Congress Cataloging-in-Publication Data

Rinehart, Susie Caldwell.
   Eliza and the dragonfly / by Susie Caldwell Rinehart ; illustrated by
Anisa Claire Hovemann.-- 1st ed.
       p. cm. -- (A sharing nature with children book)
Summary: When a dragonfly slips in and lands on her toothbrush,
Eliza escorts it to a nearby pond to learn more about these
remarkable insects.
   ISBN 1-58469-060-7 (hardback) -- ISBN 1-58469-059-3 (pbk.)
   [1. Dragonflies--Fiction.] I. Hovemann, Anisa Claire, ill. II. Title.
III. Series.
   PZ7.R468El 2004
   [E]--dc22
                                    2003021653

Dawn Publications
12402 Bitney Springs Road
Nevada City, CA 95959
800-545-7475
nature@dawnpub.com

Printed in Korea

10 9 8 7 6 5 4 3 2 1
First Edition
Design and computer production by Menagerie

There's a dragonfly in
my house. It flew in the
window and landed on
my toothbrush. I have
to tell Aunt Doris.

My Aunt Doris loves bugs.
And spiders and bees and ants.
She uses words like "magnificent"
when she sees things that sting
and soar.

Aunt Doris is an entomologist. That's somebody who studies insects. She loves to go to the pond near my house to watch bugs.

"MAGNIFICENT!" Aunt Doris says when she sees the dragonfly in my house. "This one belongs at the pond." So I hop on my bike and Aunt Doris hops on her bugcycle and we ride to the pond. The dragonfly holds on tightly. I carry it like it is the most precious gift in the world.

A great blue heron wades silently through the lilies. Aunt Doris walks along the water's edge, holding her net high like a kite. Dragonflies zoom faster than birds over the smooth surface of the pond. My toothbrush dragonfly rushes to join them.

I make a wish on every dragonfly I see. I wish I could fly as fast as a dragonfly. I wish I could ride my bike anywhere I wanted. I wish it didn't take so long to grow up.

Kersploosh! Aunt Doris slips and falls in the pond!
The great white egret pumps her wings and flies
away. A frog hops out of Aunt Doris's pocket.
Fish and leaves drip from her hair. But Aunt Doris
doesn't mind. She never minds what she looks like.
That's what I like about Aunt Doris.

Then an awful green creature the size of a paperclip swims out of her boot. She scoops it into her hand. "Eeeeewwww," I say.

"MAGNIFICENT!" Aunt Doris says.

"What is it?" I ask.

"It's a young dragonfly, Eliza. A nymph," Aunt Doris says as she lets the creature go.

"What's a dragonfly doing underwater?"

"Dragonflies begin their lives in water. When they're young, they breathe water instead of air. And they swim instead of fly. Let's make a waterscope and you can see for yourself."

Aunt Doris digs into her backpack and finds an empty juice container. She cuts off the top and the bottom. Then she takes the clean plastic wrap off her sandwich and spreads it over one end of the container. A big rubber band holds the plastic wrap in place.

"Ta Daaa! A waterscope!" Aunt Doris shows me how to look through the scope underwater.

"Wow! There's an amazing wiggling world down here!" I gasp.

"Can you see the dragonfly nymph?"

"I think so. But it doesn't look like a dragonfly. Where are its wings?" I wonder.

"Look closer," Aunt Doris says. I see four tiny wings like flower petals folded on a dragon's back.

"When will it know how to fly?" I ask.

"Eliza," Aunt Doris smiles, "A dragonfly nymph doesn't worry about when it will grow up and become a dragonfly. It doesn't wish it could fly or be more beautiful than it already is. It just mucks about in the pond, being itself. Then it wakes up one morning with wings."

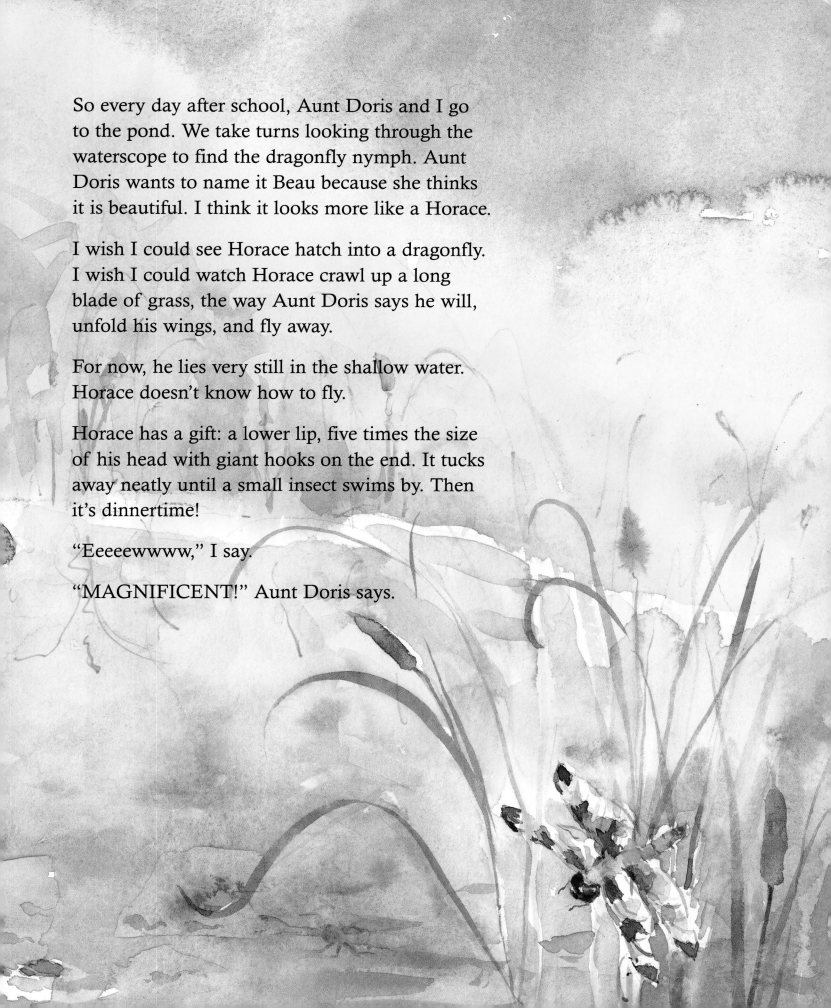

So every day after school, Aunt Doris and I go to the pond. We take turns looking through the waterscope to find the dragonfly nymph. Aunt Doris wants to name it Beau because she thinks it is beautiful. I think it looks more like a Horace.

I wish I could see Horace hatch into a dragonfly. I wish I could watch Horace crawl up a long blade of grass, the way Aunt Doris says he will, unfold his wings, and fly away.

For now, he lies very still in the shallow water. Horace doesn't know how to fly.

Horace has a gift: a lower lip, five times the size of his head with giant hooks on the end. It tucks away neatly until a small insect swims by. Then it's dinnertime!

"Eeeeewwww," I say.

"MAGNIFICENT!" Aunt Doris says.

Sometimes I bring my colored pencils and paper to the pond and draw Horace. He looks at me with half a smile on his face.

I tell Horace that soon he'll be able to breathe air instead of water. I tell him that soon he'll be able to fly faster than a swallow. And I tell him other things too. Like how I wish I was taller and had straight blond hair, and how I wish I could paint beautiful pictures that the whole world would come to see. Horace stares at me like he is trying to tell me something.

But Horace still doesn't know how to fly.

Sometimes my friends Carlos and Annie come to the pond with me after school. Horace loves an audience. He glides around on the bottom of the pond as if he were dancing. Cha-cha-cha Horace!

Carlos is teaching Horace to speak Spanish. Horace's eyes grow wide and it looks like he is listening really hard. Hola Horace!

Annie takes out her recorder and plays Horace a tune. Do-re-mi Horace!

But Horace still doesn't know how to fly.

"I'll bet he wishes he could fly," Annie says.

"Horace doesn't wish he could fly or be more beautiful than he already is," I say. Aunt Doris smiles.

"Oh no. Horace's gone!" Carlos says when he lowers the scope into the water.

"Gone? Let me look," I say. Through the waterscope I see pollywogs and fish, but no Horace. I move the scope up and down and back and forth. Still no Horace.

"Aunt Doris, where is he?"

"Here he is!" Aunt Doris waves us over excitedly. On a tall plant stem rising out of the water, Horace climbs out of the pond for the first time.

"What's happening?" asks Eliza.

"Horace is going to hatch," Aunt Doris whispers. We all lie down in the grass to watch.

At first, nothing happens. The wind swings the stem back and forth. Horace holds on like it is the only solid thing in the world.

Then he takes his first breath
of air. It moves through his
body and cracks his shell open.
Slowly, so slowly, Horace slides
out of his shell like he is
slipping out of a sleeping bag.
First his eyes appear, then his
slender body and finally his
wings. I hold my breath.
He does not move. Horace's
wings lie wrinkled across
his back.

"Is he OK?" I ask.

"Don't worry. He's pumping up and drying his wings. When it's time, he'll zoom away to eat bugs for dinner," Aunt Doris says.

"Eeeeewwww," Carlos says.

Then the sun touches Horace's eyes and we see that they are a brilliant green. The sun touches Horace's soft, wet wings and they glow like they are made of tiny stained-glass windows.

"MAGNIFICENT!" Aunt Doris says.

"Ouch!" Carlos slaps his neck and shouts, "MOSQUITOES!"

There are mosquitoes everywhere. They bite my ankles, they bite my arms, and they bite the soft places behind my knees. Carlos spins around in circles then falls down dizzy. Annie flaps her arms like she is trying to fly away. I shake my whole body like a wet dog but the mosquitoes keep whining and buzzing.

Suddenly, the buzzing stops. It is very quiet. No mosquitoes bite my knees. No mosquitoes bite my neck. There are hardly any mosquitoes at all.

What do you think happened?

"Look!" Carlos and Annie say at the same time.

A liquid-blue dragonfly with stained-glass wings silences the pond, one mosquito at a time.

"Horace!" Aunt Doris says.

"MAGNIFICENT!" I say.

Learn About Dragonflies

Common Green Darner

Dragonhunter

Have you ever noticed the way mosquitoes disappear when there's a dragonfly around? That is because dragonflies eat mosquitoes, black flies, and many other flying insects. There are over 5,000 species of dragonflies and damselflies in the world and over 430 species in North America. Horace is a Common Green Darner (Anax junius), a species that is widespread in North America.

Flying Cousins

Dragonflies and damselflies are biological cousins. Together, they are known as Odonata. To tell the difference, look at their wings. Damselflies are smaller and skinnier than dragonflies and fold their wings together over their backs when they rest, while dragonflies spread them straight out to the sides.

Emerald Spreadwing

Life Cycle

A dragonfly begins its life in water. It hatches in the summer as a dark wingless nymph and spends one to three winters on the bottom of the pond. A quiet hunter, the dragonfly nymph captures tadpoles and small insects with a powerful lower lip that extends five times beyond the length of its head. Its wings begin as two faint lines no thicker than the edge of a flower petal. As the nymph grows, it sheds its hard outer shell. A nymph may have ten to fifteen growth spurts in its life! For its final metamorphosis, a nymph will crawl out of the water for the first time, cling to a plant stem, take its first breath of air, and slide out of its shell. Then it stretches its wings open to dry in the sun and flies away for its first meal of mosquitoes. From now on, most of the dragonfly's life occurs in flight. If you're lucky, you may see a dragonfly resting on a grass stem or on your outstretched hand.

Familiar Bluet

Natural History

Did you know that a dragonfly can fly faster than many birds? It eats, mates, lays its eggs, and escapes predators at speeds of up to thirty-six miles per hour! A female dragonfly lays her eggs by repeatedly dipping the end of her body in the water or along the thin edges of underwater plants. The eggs settle to the bottom and hatch into nymphs in about a week. The nymphs mature underwater. They may live on the bottom of the pond for as short a time as six weeks, or up to five years! Out of the water, dragonflies live only for several weeks. Many dragonflies stay close to their home pond or stream but some migrate. They cannot sting humans.

Did you know that dragonflies have lived on Earth since long before dinosaurs were alive? They have been around for over 200 million years and are among the oldest living species. Even so, new species of dragonflies are still being discovered. Maybe you can find one!

Eastern Pondhawk

Wandering Glider

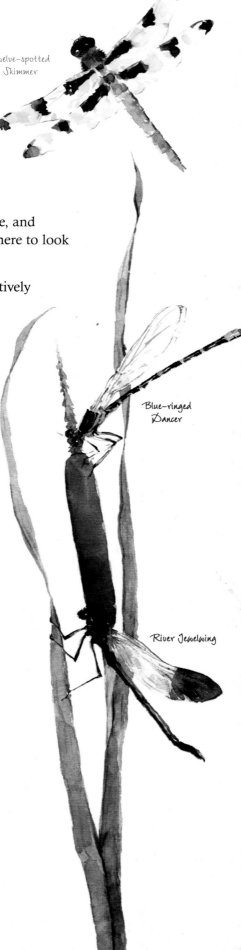
Twelve-spotted Skimmer

Blue-ringed Dancer

River Jewelwing

Resources

Read More About Dragonflies

Dragonflies Through Binoculars by Sidney Dunkle, 2000. This is a comprehensive field guide with 47 color plates of photographs and sections covering such topics as identification, habitat, biology, and conservation.

Stokes Beginner's Guide to Dragonflies and Damselflies by Blair Nikula, et al, 2002. An easy-to-use guide that helps you find, observe, and appreciate over 100 species of Odonata with excellent full-color photographs, tips on where to look for dragonflies and damselflies and fascinating details about their behavior.

What's in the Pond? Anne Hunter, 1999. A small formatted book, well-written and attractively illustrated that would be useful for field trips with young children.

A Few Online Resources Worth Exploring

Dragonfly Society of the Americas
2091 Partridge Lane
Binghamton, NY 13903
www.afn.org/~iori/dsaintro.html

Organized in 1988 to develop better communication between amateur and professional dragonfly enthusiasts, DSA encourages scientific research and habitat preservation.

Odonata Information Network
www.afn.org/~iori

The home base for all dragonfly and damselfly enthusiasts. Their website is the perfect place to post questions and information requests, or to find out more about other organizations in the Americas. The site includes a lengthy, updated list of links to other Odonata sites.

Saffron-winged Meadowhawk

Ode News
www.odenews.net

An on-line newsletter about dragonflies and damselflies in southern New England. This user-friendly site provides excellent information regarding books, organizations, and research efforts related to dragonflies and damselflies currently occurring in North America.

California On-line Dragonfly Guide Site
http://www.sonic.net/dragonfly

Here you will find links not only to information on and photos of all the species of dragonflies and damselflies in California, but also information on books and general information about dragonflies.

Other Regional Websites

There are many other websites about dragonflies in particular regions. If you enter the name of your state or province plus "dragonfly" or "dragonflies" in a search engine, you may find a site specific to where you live.

When Susie Caldwell Rinehart is not busy mucking about in the pond watching dragonflies, she is a teacher and writer. Born in Toronto, Canada, she has degrees in English and Science. Susie believes that the best kind of learning takes place on the edge between science and art. *Eliza and the Dragonfly* is her first children's book. Susie lives in Vershire, Vermont with her husband Kurt and their son Cole.

It seems that Anisa Claire Hovemann has always been an artist. As a three year-old, she would form remarkably lifelike dinosaurs out of her colorful playdough. At the Waldorf elementary school she attended, her artistic tendencies fit naturally into the curriculum. She is currently a senior at the Maryland Institute College of Art, pursuing a degree in fine arts—but took off a semester to finish this project, her first illustrated book. Her research included spending quite a bit of time at "Dragonfly Pond" near her home in Nevada City, California.

A Few Other Nature Awareness Books from Dawn Publications

Under One Rock: Bugs, Slugs and other Ughs by Anthony Fredericks, illustrated by Jennifer DiRubbio. No child will be able to resist looking under a rock after reading this rhythmic, engaging story.

Girls Who Looked Under Rocks, by Jeannine Atkins, illustrated by Paula Connor. Six girls, from the 17th to the 20th century, didn't run from spiders or snakes but crouched down to take a closer look. They became pioneering naturalists, passionate scientists, and energetic writers or artists.

Earth Day Birthday by Pattie Schnetzler, illustrated by Chad Wallace. To the tune of "The Twelve Days of Christmas," here is a sing-along, read-along book that honors the animals, the environment, and a universal holiday all in one fresh approach.

Sunshine On My Shoulders by John Denver, adapted and illustrated by Christopher Canyon. This heartwarming adaptation of a simple, sweet song reminds one of fresh and free childhood days. The first in a series adapting some of Denver's environmentally-aware songs for children.

Do Animals Have Feelings, Too? by David Rice, illustrated by Trudy Calvert, presents fascinating true stories of animal behavior, and asks the reader whether they think the animals' actions show feelings or instinct.

Dawn Publications is dedicated to inspiring in children a deeper understanding and appreciation for all life on Earth. To view our full list of titles or to order, please visit our web site at www.dawnpub.com, or call 800-545-7475.